The Genomic Revolution: Unlocking the Secrets of Life for Everyone

Kyler Lane

Copyright © [2023]

Title: The Genomic Revolution: Unlocking the Secrets of Life for Everyone
Author's: Kyler Lane

This book was printed and published by [Publisher's: **Kyler Lane**] in [2023]

ISBN:

TABLE OF CONTENT

Chapter 1: Introduction to Functional Genomics   07

Understanding Genomics and Functional Genomics

The Importance of Functional Genomics in Everyday Life

Historical Overview of Functional Genomics Research

Chapter 2: The Basics of Genomics   14

DNA Structure and Function

Genes and Genomes

Genetic Variation and Mutation

Chapter 3: Functional Genomics Techniques   20

Transcriptomics: Studying Gene Expression

Proteomics: Analyzing Protein Structures and Functions

Metabolomics: Investigating Metabolic Pathways

Epigenomics: Exploring Epigenetic Modifications

Chapter 7: Accessible Functional Genomics for Everyone 48

Genomic Literacy and Education

Open Access Data and Resources

Citizen Science and Participatory Genomics

Chapter 8: Challenges and Limitations in Functional Genomics 54

Data Interpretation and Analysis

Technical and Financial Constraints

Ethical and Regulatory Challenges

Chapter 9: Case Studies in Functional Genomics 60

Unraveling Genetic Causes of Complex Diseases

Leveraging Functional Genomics for Precision Agriculture

Conservation Genomics: Protecting Endangered Species

Chapter 10: Conclusion 67

Recap of Key Concepts

The Impact of Functional Genomics on Society

Looking Ahead: What Lies Beyond the Genomic Revolution?

Chapter 1: Introduction to Functional Genomics

Understanding Genomics and Functional Genomics

Genomics is a field of study that has revolutionized our understanding of life and its intricate workings. In this subchapter, we will delve into the fascinating world of genomics and functional genomics, and explore how these disciplines are unlocking the secrets of life for everyone.

Genomics is the study of an organism's complete set of DNA, including all of its genes. It involves the analysis of the structure, function, and evolution of genes and their interactions within a genome. By uncovering the genetic material that makes up an organism, genomics provides crucial insights into how traits are inherited and how they influence an organism's development and overall health.

Functional genomics, on the other hand, focuses on understanding how genes function and how they interact with each other and the environment. It aims to decipher the complex network of gene interactions that give rise to the diverse properties of living organisms. By studying the function of genes and their products, functional genomics sheds light on the underlying mechanisms of diseases and helps develop targeted therapies and personalized medicine.

Genomics and functional genomics have far-reaching applications in various fields, including medicine, agriculture, and environmental science. By identifying genetic variations associated with diseases, genomics has revolutionized the diagnosis and treatment of genetic disorders. It has also played a pivotal role in the development of precision medicine, enabling healthcare professionals to tailor treatments based on an individual's genetic makeup.

In agriculture, genomics has transformed the breeding of crops and livestock. By identifying desirable genetic traits, such as disease resistance and increased yield, scientists can develop improved varieties through selective breeding and genetic engineering. This has the potential to enhance food security and sustainability, ensuring a brighter future for our planet.

Furthermore, genomics has opened up new avenues for environmental science by enabling the study of organisms' interactions with their environment. By analyzing the genomes of organisms in different ecosystems, scientists can better understand the impact of human activities on biodiversity and ecosystem health. This knowledge is crucial for developing effective conservation strategies and mitigating the environmental challenges we face today.

In conclusion, genomics and functional genomics have transformed our understanding of life and its complexities. The insights gained from these disciplines have revolutionized medicine, agriculture, and environmental science, benefiting not only scientists and researchers but also the general public. Whether you are interested in genetics and genomics or simply curious about the secrets of life, exploring the world of genomics will undoubtedly be an enlightening and awe-inspiring journey.

The Importance of Functional Genomics in Everyday Life

In today's world, where technological advancements have become an integral part of our lives, functional genomics plays a crucial role in unraveling the mysteries of life. The study of functional genomics has revolutionized the field of genetics and genomics, leading to a deeper understanding of how genes function and interact with each other. This subchapter aims to highlight the significance of functional genomics and its relevance to everyone, regardless of their background or interests.

Functional genomics investigates how genes work together to carry out specific functions within an organism. It delves into the intricate mechanisms underlying genetic processes, such as gene expression, protein synthesis, and cellular pathways. By studying how genes function in different contexts, functional genomics provides valuable insights into the development, maintenance, and regulation of living organisms.

One of the most impactful applications of functional genomics is in the field of medicine. It has paved the way for personalized medicine, where treatments are tailored to an individual's genetic makeup. Functional genomics allows scientists to identify genetic variations that may predispose individuals to certain diseases or affect their response to medications. This knowledge enables healthcare professionals to make informed decisions about treatment plans, improving patient outcomes and reducing adverse reactions.

Moreover, functional genomics has revolutionized agriculture and food production. By understanding the genes responsible for desirable traits in crops and livestock, scientists can develop genetically modified organisms (GMOs) that are more resistant to diseases, pests, and environmental stressors. This advancement ensures food security, increases crop yield, and enhances nutritional value, ultimately benefiting everyone by providing a sustainable and abundant food supply.

Functional genomics also plays a crucial role in environmental conservation and sustainable development. By studying the genes of different species, scientists can gain insights into their adaptations, population dynamics, and ecological interactions. This knowledge aids in the conservation of endangered species, the management of ecosystems, and the mitigation of climate change impacts. Functional genomics thus contributes to the preservation of biodiversity and the overall health of our planet.

In addition to these practical applications, functional genomics has profound implications for our understanding of human history, evolution, and the fundamental principles of life. It allows us to trace our genetic ancestry, unravel the origins of diseases, and explore the complexities of human physiology. Functional genomics fuels scientific curiosity and encourages interdisciplinary collaborations, fostering innovation and driving progress in various fields.

In conclusion, functional genomics is a powerful tool that has far-reaching implications for everyone, regardless of their background or interests. From personalized medicine to sustainable agriculture and environmental conservation, functional genomics has the potential to improve our lives, protect our planet, and unlock the secrets of life itself. By embracing the genomic revolution, we can harness the power of functional genomics and pave the way for a brighter and more informed future.

Historical Overview of Functional Genomics Research

In the ever-evolving field of genetics and genomics, functional genomics research has played a pivotal role in unraveling the intricate secrets of life. This subchapter will take you on a journey through time, exploring the fascinating historical milestones that have shaped this revolutionary field.

Functional genomics research aims to understand the function and interactions of genes within an organism, providing insights into the complex mechanisms that govern life processes. The roots of this discipline can be traced back to the early 20th century when scientists first began to unravel the mysteries of genetics.

One of the key breakthroughs in functional genomics research came in the 1950s with the discovery of the structure of DNA by James Watson and Francis Crick. This groundbreaking discovery laid the foundation for understanding how genes encode the instructions for building and maintaining living organisms.

As technology advanced, the field of functional genomics underwent a major transformation in the 1980s with the development of DNA sequencing techniques. This allowed researchers to unravel the genetic code and paved the way for studying genes on a larger scale.

The introduction of high-throughput sequencing technologies in the 1990s revolutionized functional genomics research. These techniques enabled scientists to sequence entire genomes and explore the vast landscape of genes present in organisms. The Human Genome Project, completed in 2003, marked a significant milestone in this regard, providing a comprehensive map of the human genome.

With the advent of next-generation sequencing technologies in the early 2000s, functional genomics research entered a new era. Scientists could now study gene expression patterns, protein interactions, and regulatory networks in unprecedented detail. This wealth of information has allowed researchers to dive deeper into the complexities of gene function and unravel the intricate web of interactions that govern various biological processes.

Today, functional genomics research continues to push boundaries and uncover new insights into the mechanisms of life. It plays a crucial role in disease research, drug development, and personalized medicine, offering hope for improved diagnostics and targeted therapies.

As we move forward, the field of functional genomics holds immense promise for understanding the secrets of life and transforming healthcare. By unraveling the mysteries of our genetic makeup, functional genomics research is unlocking a new era of possibilities for everyone, paving the way for a brighter and healthier future.

Chapter 2: The Basics of Genomics

DNA Structure and Function

In order to understand the fascinating world of genetics and genomics, it is crucial to grasp the basics of DNA structure and its fundamental role in the functioning of life. DNA, short for deoxyribonucleic acid, is the blueprint of life and the molecule that holds the key to our genetic inheritance.

At its core, DNA is a long, twisted ladder-like structure, known as the double helix. This elegant structure was first discovered by James Watson and Francis Crick in 1953, marking a groundbreaking moment in the history of science. The double helix consists of two intertwined strands, each made up of a series of building blocks called nucleotides. These nucleotides are composed of a sugar molecule, a phosphate group, and one of four nitrogenous bases: adenine (A), guanine (G), cytosine (C), and thymine (T).

The arrangement of these nitrogenous bases is what determines our unique genetic code. Adenine pairs with thymine, and guanine pairs with cytosine, forming complementary base pairs. This pairing allows DNA to replicate itself accurately during cell division, ensuring the transmission of genetic information from one generation to the next.

Beyond its role as a static genetic code, DNA also serves as a dynamic molecule with multiple functions. It acts as a template for the synthesis of another type of nucleic acid called ribonucleic acid (RNA). RNA plays essential roles in translating genetic information into proteins, which are the building blocks of life. Through a process known as transcription, DNA is used as a template to produce messenger RNA (mRNA), which carries the genetic instructions from the nucleus to the protein-making machinery in the cell.

Additionally, DNA is not confined to the nucleus of a cell. Mitochondria, the energy-producing powerhouses of our cells, also contain their own DNA. This mitochondrial DNA (mtDNA) is inherited exclusively from the mother and plays a vital role in energy production and other cellular processes.

Understanding the structure and function of DNA is crucial for comprehending the complexities of genetics and genomics. It allows us to explore the fascinating world of inherited traits, genetic diseases, and the potential for personalized medicine. Whether you are a student, a healthcare professional, or simply curious about the secrets of life, delving into the remarkable world of DNA will unlock a new level of appreciation for the genomic revolution that is shaping our understanding of ourselves and all living organisms.

Genes and Genomes

In the vast landscape of genetics and genomics, the study of genes and genomes reveals the intricate blueprint of life itself. Understanding the fundamental building blocks of our existence opens up a world of possibilities, from unraveling the mysteries of inherited diseases to shaping the future of personalized medicine. In this subchapter, we embark on a journey to explore the fascinating realm of genes and genomes, where the secrets of life are waiting to be unlocked.

Genes are the instructions encoded within our DNA that determine our traits, characteristics, and even susceptibilities to certain diseases. They are the functional units of heredity, passed down from one generation to the next. Genomes, on the other hand, encompass all the genetic material within an organism, including both the coding and non-coding regions of DNA. They hold the key to understanding the complex interplay between genes and their environment.

Advancements in technology, such as next-generation sequencing, have revolutionized the field of genomics. We now have the ability to read and analyze entire genomes, providing us with unprecedented insights into the genetic makeup of organisms. This wealth of information has paved the way for breakthroughs in medicine, agriculture, and even forensic science.

One of the most significant applications of genomics lies in the field of personalized medicine. By analyzing an individual's genome, scientists can identify genetic variations that may influence their response to certain drugs or their predisposition to certain diseases. This allows for tailored treatment plans and preventive measures, ultimately improving patient outcomes.

Genes and genomes also play a crucial role in understanding the mechanisms behind genetic disorders. By studying the specific genes involved, scientists can unravel the underlying causes of these conditions and develop targeted therapies. Additionally, genomics has shed light on the complex interactions between genes and the environment, helping us understand how lifestyle factors and external stimuli can influence gene expression and overall health.

In conclusion, the study of genes and genomes holds immense promise in unlocking the secrets of life for everyone. Whether you are fascinated by the science behind genetics or simply curious about how genomics impacts our lives, delving into the world of genes and genomes opens up a world of knowledge and possibilities. From personalized medicine to unraveling the mysteries of inherited diseases, the genomic revolution is transforming the way we approach healthcare and our understanding of ourselves.

Genetic Variation and Mutation

In the vast realm of genetics and genomics, one of the most fascinating aspects is the concept of genetic variation and mutation. These fundamental processes lie at the core of life, shaping the diversity we see in all living organisms. Understanding genetic variation and mutation is crucial to unraveling the secrets of life, and this subchapter will delve into their significance and implications.

Genetic variation refers to the diversity of genetic material within a population or species. It arises from various sources, such as genetic recombination during sexual reproduction and the accumulation of mutations over time. This variation is what gives rise to the unique traits and characteristics that make each individual distinct.

Mutation, on the other hand, is a spontaneous change in the DNA sequence. It can occur due to errors during DNA replication, exposure to mutagens like radiation or chemicals, or even random chance. While some mutations can have detrimental effects on an organism's health or viability, others can be neutral or even beneficial. In fact, mutations are the driving force behind evolution, allowing species to adapt to changing environments over generations.

The study of genetic variation and mutation has far-reaching implications. It helps us understand the genetic basis of various diseases, as well as how different individuals respond to treatments. Genetic variation also plays a crucial role in personalized medicine, as it influences an individual's susceptibility to diseases and their response to specific drugs. By uncovering the links between genetic variation and disease, we can develop targeted therapies and interventions that improve patient outcomes.

Furthermore, genetic variation and mutation have profound implications for biodiversity and conservation efforts. The greater the genetic variation within a species, the better its chances of survival in the face of environmental challenges. Understanding the genetic diversity of endangered species allows us to develop strategies to preserve their unique genetic heritage and prevent their extinction.

In conclusion, genetic variation and mutation are the building blocks of life and crucial components of genetics and genomics. They shape the diversity we see in all living organisms and underpin our understanding of disease, personalized medicine, and conservation. By unlocking the secrets of genetic variation and mutation, we can unlock the secrets of life itself.

Chapter 3: Functional Genomics Techniques

Transcriptomics: Studying Gene Expression

In the world of genetics and genomics, one of the most fascinating areas of research is transcriptomics. It is a field that delves into the study of gene expression, unraveling the intricate mechanisms that govern how our genes are turned on or off.

Genes are the fundamental units of heredity, carrying the instructions necessary for the development and functioning of all living organisms. However, not all genes are active at all times. Transcriptomics aims to understand the complex processes that regulate gene expression, shedding light on how our genetic information is translated into functional molecules.

At its core, gene expression is the process by which information encoded in our DNA is transcribed and translated into proteins. Transcriptomics focuses on the first part of this process, investigating the transcripts or RNA molecules that are produced when genes are activated. These RNA molecules play crucial roles as intermediaries between DNA and proteins, acting as messengers that carry genetic information to the protein synthesis machinery.

Transcriptomics utilizes advanced technologies to analyze the abundance, diversity, and activity of RNA molecules in a given cell or tissue. One of the main tools used in this field is high-throughput sequencing, which enables the simultaneous analysis of thousands or even millions of RNA molecules. This powerful technique allows scientists to identify which genes are active in a specific biological context and to quantify their expression levels.

By studying gene expression patterns, transcriptomics provides valuable insights into various biological processes. It helps us understand how different cells in our bodies acquire distinct identities, how diseases arise due to dysregulation of gene expression, and how environmental factors can influence gene activity.

Furthermore, transcriptomics has practical applications in medicine and biotechnology. It can help identify potential drug targets, predict patient responses to treatment, and discover biomarkers for various diseases. By deciphering the intricacies of gene expression, transcriptomics is revolutionizing our understanding of human health and disease.

In conclusion, transcriptomics is a powerful tool that allows us to explore the fascinating world of gene expression. It provides a deeper understanding of how our genes work, how they are regulated, and how they contribute to the complexity of life. With its potential to uncover the secrets of life at the molecular level, transcriptomics is shaping the future of genetics and genomics, bringing us closer to unlocking the mysteries of our own existence.

Proteomics: Analyzing Protein Structures and Functions

In the world of genetics and genomics, understanding the intricate workings of proteins is crucial. Proteomics, the study of proteins and their functions, plays a pivotal role in unraveling the mysteries of life. In this subchapter, we will delve into the fascinating realm of proteomics and explore how it contributes to our understanding of biological systems.

Proteins are the building blocks of life, carrying out essential functions within our cells. They are responsible for everything from catalyzing biochemical reactions to transporting molecules and coordinating cellular processes. The structure and function of proteins are intricately linked, and proteomics aims to decipher this complex relationship.

One of the primary goals of proteomics is to identify and characterize all the proteins in a given organism, known as the proteome. By doing so, scientists can gain insights into the diverse array of proteins present and understand how they interact with one another to maintain cellular homeostasis. This holistic approach provides a comprehensive view of the intricate networks within an organism, enabling scientists to uncover vital biological pathways and potential therapeutic targets.

Proteomics employs a range of sophisticated techniques to analyze protein structures and functions. These include mass spectrometry, which allows for the identification and quantification of proteins, as well as X-ray crystallography and nuclear magnetic resonance (NMR) spectroscopy, which provide detailed information about protein structures. Additionally, advanced computational tools are utilized to predict protein structures and analyze their interactions with other molecules.

The applications of proteomics are vast and impact various fields, including medicine, agriculture, and environmental science. In the medical field, proteomics has revolutionized the diagnosis and treatment of diseases. By analyzing the proteome of a patient, doctors can identify specific biomarkers associated with diseases and develop personalized treatment strategies. Proteomics also plays a crucial role in drug discovery, enabling scientists to design targeted therapies that interact with specific proteins involved in disease processes.

In agriculture, proteomics helps enhance crop yield and quality by identifying proteins responsible for important traits such as disease resistance and nutritional content. Furthermore, proteomics contributes to environmental science by monitoring protein expression in organisms exposed to pollutants, providing insights into their response to environmental stressors.

Proteomics is a rapidly evolving field with immense potential for unlocking the secrets of life. By analyzing protein structures and functions, scientists are unraveling the intricate mechanisms that underlie biological systems. Whether you are a genetics enthusiast or simply curious about the inner workings of life, understanding the power of proteomics will undoubtedly ignite your fascination with the genomic revolution.

Metabolomics: Investigating Metabolic Pathways

Metabolomics is an exciting and rapidly evolving field in the realm of genetics and genomics. It involves the study of small molecules called metabolites, which are the end products of cellular processes. By analyzing these metabolites, scientists can gain valuable insights into the metabolic pathways that occur within our cells.

Metabolic pathways are a series of chemical reactions that take place in our cells to produce energy, synthesize new molecules, and eliminate waste products. These pathways are like intricate roadmaps that guide the flow of molecules and energy throughout our bodies. Understanding them is crucial for unlocking the secrets of life.

In the past, geneticists focused primarily on studying genes and their role in disease and development. However, as the field of genomics has advanced, researchers have realized that genes alone cannot tell the whole story. Metabolomics offers a holistic approach, bridging the gap between genes and their ultimate impact on our health.

One of the key goals of metabolomics is to identify and quantify metabolites present in a given biological sample. This is done using a combination of analytical techniques such as mass spectrometry and nuclear magnetic resonance spectroscopy. By comparing the metabolite profiles of healthy individuals with those affected by diseases, scientists can identify biomarkers that can help diagnose and monitor various conditions.

Metabolomics has already made significant contributions to fields like personalized medicine, nutrition, and drug development. For instance, by analyzing metabolites in a patient's blood or urine, doctors can tailor treatment plans to their specific metabolic profile, increasing the chances of successful outcomes. Similarly, researchers can use metabolomics to identify potential drug targets and develop more effective therapies.

Moreover, metabolomics can shed light on how our bodies respond to different diets and lifestyle choices. By studying the metabolic pathways affected by certain foods or exercise regimens, we can optimize our health and well-being. This knowledge can also help in designing personalized nutrition plans for individuals, improving their overall quality of life.

In conclusion, metabolomics is a powerful tool that allows us to investigate the intricate metabolic pathways within our cells. By studying the small molecules produced during these processes, we can gain a deeper understanding of how our genes and environment interact to shape our health and well-being. As the field continues to advance, metabolomics promises to revolutionize medicine, nutrition, and drug development, ultimately unlocking the secrets of life for everyone.

Epigenomics: Exploring Epigenetic Modifications

In recent years, the field of genetics and genomics has witnessed a groundbreaking revolution that has transformed our understanding of life itself. This revolution has brought forth a new frontier of research known as epigenomics, which delves into the intricate world of epigenetic modifications. Epigenomics explores how our genes are regulated and influenced by factors beyond the DNA sequence, shedding light on the complex interplay between genetics and the environment.

Epigenetic modifications refer to changes in gene expression that occur without altering the underlying DNA sequence. These modifications can be influenced by a variety of factors, such as environmental exposures, lifestyle choices, and even our thoughts and emotions. They act as molecular switches, turning genes on or off, and play a crucial role in development, aging, and disease susceptibility.

Understanding epigenetic modifications opens up new avenues for research and potential interventions in various fields, including medicine, agriculture, and environmental sciences. It allows scientists to delve deeper into the mechanisms behind diseases, uncovering potential therapeutic targets and developing personalized treatments. In the realm of cancer research, epigenomics has revealed the crucial role of these modifications in the development and progression of tumors, paving the way for targeted therapies that can reverse aberrant epigenetic patterns.

Moreover, epigenomics has shed light on the impact of our environment on gene expression. It has helped unravel the relationship between nutrition, stress, and disease susceptibility, emphasizing the importance of a healthy lifestyle in maintaining optimal gene function. This newfound knowledge has inspired individuals to take charge of their health and make informed choices that can positively influence their genetic expression.

Epigenomics also holds promise in the field of agriculture, where it can aid in crop improvement and environmental sustainability. By understanding the epigenetic modifications that occur in response to different environmental conditions, researchers can develop crops that are more resilient to climate change or enhance the nutritional content of food.

In conclusion, epigenomics represents a fascinating frontier in the field of genetics and genomics. It offers a deeper understanding of how our genes are regulated and influenced by the environment, giving us new insights into development, aging, and disease. By unraveling the complexities of epigenetic modifications, we are unlocking the secrets of life and paving the way for a future where personalized medicine, sustainable agriculture, and healthier lifestyles become the norm for everyone.

Chapter 4: Applications of Functional Genomics

Disease Diagnosis and Personalized Medicine

In the ever-evolving field of genetics and genomics, one of the most exciting and promising areas is disease diagnosis and personalized medicine. Thanks to groundbreaking advancements and the power of genomic sequencing, we are now able to unlock the secrets of life and tailor medical treatments to individual patients like never before.

Traditionally, the diagnosis of diseases has relied on symptoms, medical history, and physical examinations. However, with the advent of genomic technology, we now have the ability to delve deep into an individual's genetic makeup to identify the root causes of diseases. By analyzing an individual's DNA, scientists can pinpoint genetic mutations and variations that may increase the risk of developing certain diseases.

This newfound knowledge opens up a world of possibilities for personalized medicine. Rather than providing one-size-fits-all treatments, doctors can now tailor therapies to a patient's unique genetic profile. This approach allows for more targeted and effective treatments, minimizing potential side effects and improving overall patient outcomes.

One of the most notable examples of personalized medicine is in the field of cancer treatment. By analyzing the genetic mutations present in a tumor, oncologists can identify specific targeted therapies that are most likely to be effective for a particular patient. This approach has revolutionized cancer care, leading to improved survival rates and a higher quality of life for many patients.

Beyond cancer, personalized medicine holds promise for a wide range of other diseases, including cardiovascular disorders, neurological conditions, and rare genetic disorders. By understanding the genetic basis of these diseases, researchers can develop innovative treatments that address the underlying causes rather than just managing symptoms.

However, it is important to note that personalized medicine is still in its early stages, and there are many challenges to overcome. The cost of genomic sequencing, the interpretation of vast amounts of genetic data, and the ethical considerations surrounding genetic testing are just a few of the hurdles that need to be addressed. Nonetheless, the potential benefits are immense, and the field continues to progress at a rapid pace.

In conclusion, disease diagnosis and personalized medicine are at the forefront of the genomic revolution. By harnessing the power of genetics and genomics, we can better understand the causes of diseases and tailor treatments to individual patients. While there are still obstacles to overcome, the promise of personalized medicine offers hope for a future where healthcare is truly personalized, effective, and accessible to everyone.

Drug Discovery and Development

In this subchapter, we delve into the fascinating world of drug discovery and development, where the secrets of life are harnessed to improve the health and well-being of everyone. The field of genetics and genomics plays a pivotal role in this process, revolutionizing how we approach the discovery and development of life-saving medications.

Drug discovery is a complex and meticulous process that involves identifying potential targets for treatment, designing and synthesizing compounds, and thoroughly testing their safety and efficacy. Thanks to advancements in genetics and genomics, scientists now have a powerful toolkit to accelerate this process and develop more precise and effective therapies.

Genomics, the study of an organism's entire set of genes, has provided invaluable insights into the genetic basis of diseases. Researchers can now identify specific genes or mutations that contribute to the development of a disease, allowing them to target these genetic abnormalities directly. This knowledge has sparked a new wave of targeted therapies, tailored to the individual patient's genetic makeup.

Furthermore, genetics plays a crucial role in drug development by helping us understand how individuals respond differently to medications. Through the field of pharmacogenomics, scientists are uncovering how genetic variations can influence an individual's response to drugs, including their efficacy and potential side effects. This knowledge allows for personalized medicine, where treatments can be tailored to an individual's genetic profile, maximizing effectiveness and minimizing adverse reactions.

Advances in genetic engineering and gene editing technologies, such as CRISPR-Cas9, offer even greater potential for drug discovery. Scientists can now directly modify genes, creating precise changes that may lead to the development of new therapies or the correction of genetic defects.

However, drug discovery and development is a rigorous and time-consuming process, requiring extensive testing and regulatory approval. It often takes years and significant financial investments before a new drug reaches the market. Nonetheless, the advancements in genetics and genomics have accelerated this process, bringing hope for faster and more effective treatments.

In conclusion, the field of drug discovery and development is undergoing a genomic revolution. The understanding of genetics and genomics has transformed how we approach the development of new medications, allowing for targeted therapies and personalized medicine. As these advancements continue to unfold, the future of healthcare is bright, offering hope for improved treatments and better outcomes for everyone.

Agricultural Improvements through Functional Genomics

In recent years, the field of functional genomics has made significant strides in revolutionizing the agricultural industry. Through the study of how genes function and interact within organisms, scientists have been able to unlock a wealth of information that holds immense potential for improving crop yield, resistance to diseases, and overall agricultural sustainability.

Functional genomics allows researchers to identify and analyze the functions of genes involved in various agricultural traits. By understanding the specific genes responsible for desirable traits such as drought tolerance, disease resistance, and increased nutrient content, scientists can now develop innovative breeding techniques and genetically modify crops to enhance these characteristics.

One of the major breakthroughs in agricultural improvements through functional genomics has been the development of genetically modified organisms (GMOs). By introducing specific genes into crop plants, scientists have been able to confer resistance to pests and diseases, reducing the need for harmful chemical pesticides. GMOs have also shown promise in enhancing crop productivity, ensuring food security for a growing global population.

Another significant application of functional genomics in agriculture is the development of marker-assisted selection (MAS). MAS allows breeders to select for specific traits by identifying genetic markers associated with those traits. This technique significantly speeds up the traditional breeding process, enabling the development of new crop varieties with improved yield, quality, and resistance to environmental stressors.

Functional genomics has also played a crucial role in enhancing plant nutrition. By identifying the genes responsible for nutrient uptake and metabolism, scientists have been able to develop crops with enhanced nutritional content. This has the potential to address malnutrition and nutrient deficiencies in regions where access to a diverse diet is limited.

Furthermore, functional genomics has opened up new avenues for precision agriculture. By understanding the genetic basis of plant responses to environmental factors such as temperature, water availability, and soil conditions, farmers can optimize resource management and reduce environmental impacts. This knowledge allows for targeted interventions, such as adjusting irrigation schedules or applying specific fertilizers, resulting in more efficient and sustainable agricultural practices.

The advancements in functional genomics have undoubtedly revolutionized the agricultural industry, offering immense opportunities for improved crop yield, enhanced nutritional content, and sustainable farming practices. As our understanding of genes and their functions continues to expand, the potential for further agricultural improvements through functional genomics is limitless. By harnessing the power of genomics, we can pave the way for a future of food security, environmental sustainability, and improved livelihoods for farmers worldwide.

Environmental Impacts and Conservation Efforts

In the quest to unravel the mysteries of life, the field of genetics and genomics has made tremendous strides. However, with these advancements come significant environmental impacts that cannot be ignored. As we delve deeper into the secrets of life, it becomes crucial to understand and address the consequences our actions have on the environment. This subchapter aims to shed light on the environmental impacts of genetic and genomic research while highlighting the ongoing conservation efforts to mitigate them.

Genetic and genomic research often requires extensive laboratory work, which consumes energy and produces waste. The energy demands of DNA sequencing machines and other genetic analysis equipment contribute to greenhouse gas emissions, exacerbating climate change. Additionally, the disposal of hazardous chemicals and biological materials used in these experiments poses a risk to both human health and the environment.

However, the scientific community recognizes its responsibility to mitigate these environmental impacts by adopting more sustainable practices. Many research institutions are actively implementing energy-efficient measures, such as using renewable energy sources and optimizing laboratory equipment to reduce energy consumption. Proper waste management protocols are being established to minimize the release of harmful substances into the environment.

Conservation efforts are also underway to protect the biodiversity that genetics and genomics aim to unravel. By understanding the genetic makeup of various species, scientists can better assess their vulnerability to environmental change and develop conservation strategies accordingly. Genomic approaches help identify endangered populations, track their genetic diversity, and inform breeding programs to prevent genetic bottlenecks.

Furthermore, genetics and genomics have played a crucial role in wildlife conservation. DNA analysis techniques have been instrumental in combating illegal wildlife trade and identifying poached animal products, aiding law enforcement efforts. By understanding the genetic diversity within species, scientists can prioritize conservation efforts and ensure the survival of endangered populations.

Education and public awareness are essential components of conservation efforts. By disseminating knowledge about the environmental impacts of genetic and genomic research, we can foster a sense of responsibility among individuals and communities. Encouraging sustainable practices, such as reducing energy consumption and promoting recycling, can make a significant impact when adopted collectively.

In conclusion, the genomic revolution has brought us closer to unlocking the secrets of life, but it also comes with environmental consequences. However, through dedicated conservation efforts and sustainable practices, we can mitigate these impacts and ensure that the benefits of genetics and genomics are not overshadowed by their ecological footprint. By understanding the importance of environmental conservation, we can collectively work towards a more sustainable future where scientific progress goes hand in hand with protecting the natural world.

Chapter 5: Ethical Considerations in Functional Genomics

Privacy and Genetic Information

In the age of the genomic revolution, where advancements in genetics and genomics are unlocking the secrets of life, the issue of privacy with regards to genetic information has become a pressing concern. As our understanding of the human genome deepens, so does the potential for misuse and discrimination based on this intimate knowledge of our genetic makeup. This subchapter aims to shed light on the intricacies of privacy in relation to genetic information, and why it is a matter that concerns everyone.

Genetic information is unique and incredibly personal. It contains a wealth of data about our health, susceptibility to diseases, and even our ancestry. However, this very uniqueness also makes it vulnerable to misuse. The unauthorized access to genetic data can lead to discrimination in various forms, such as denial of employment, insurance coverage, or even stigmatization within society. Therefore, it is crucial to establish robust safeguards to protect the privacy of genetic information.

One of the key concerns regarding privacy and genetic information is the potential for misuse by insurance companies. The fear of being denied coverage or facing exorbitant premiums due to genetic predispositions can dissuade individuals from seeking genetic testing, even when it could provide valuable insights into their health. To address this issue, regulations must be put in place to prevent insurers from using genetic information against individuals and ensure that access to insurance remains fair and unbiased.

Another aspect of privacy is the potential for data breaches and unauthorized access to genetic databases. As the field of genetics and genomics advances, large-scale databases are being created to store genetic information. These databases are invaluable for research and medical advancements, but they also pose risks to privacy. It is essential to establish robust security measures to protect these databases and ensure that individuals' genetic information remains confidential.

Moreover, privacy concerns extend beyond individuals. Families and communities can also be affected by the disclosure of genetic information. The revelation of certain genetic conditions within a family can have far-reaching consequences, impacting not only the individual but also their relatives. Therefore, it is crucial to establish guidelines and regulations that allow individuals to maintain control over who has access to their genetic information and who can make decisions on their behalf.

In conclusion, privacy and genetic information are intertwined in the genomic revolution. As the field of genetics and genomics continues to advance, it is crucial to address privacy concerns to ensure that individuals can benefit from the knowledge of their genetic makeup without facing discrimination or loss of control over their information. By establishing regulations, safeguarding genetic databases, and promoting awareness, we can navigate the genomic revolution while protecting the privacy and dignity of every individual.

Genetic Discrimination and Social Implications

In the rapidly advancing field of genetics and genomics, it is not just our understanding of life that is evolving; it is also our comprehension of the social implications that come with this newfound knowledge. Genetic discrimination is one such consequence that has gained significant attention in recent years. This subchapter aims to explore the concept of genetic discrimination and its impact on individuals and society as a whole.

Genetic discrimination can be defined as the differential treatment of individuals based on their genetic information. With advancements in technology, it is now possible to uncover an individual's predisposition to certain diseases, their ancestry, and even their psychological traits through genetic testing. While this knowledge can be empowering, it also raises concerns about the potential misuse and abuse of genetic information.

One of the most significant social implications of genetic discrimination is its impact on personal autonomy and privacy. Individuals may fear that their genetic information could be used against them by employers, insurance companies, or even in legal proceedings. This fear can lead to individuals being hesitant about seeking genetic testing or sharing their genetic information, ultimately hindering scientific progress and the potential for personalized medicine.

Moreover, genetic discrimination can perpetuate existing social inequalities. Certain genetic traits or predispositions may be unfairly stigmatized, leading to discrimination in employment, housing, or access to healthcare. This discrimination can disproportionately affect marginalized communities, exacerbating existing disparities and creating new ones. It is essential that we address these social inequalities and strive for a future where genetic information is not a basis for discrimination or prejudice.

To mitigate the negative consequences of genetic discrimination, legislation and policies need to be put in place to protect individuals' genetic privacy and prevent discrimination. These laws should ensure that genetic information is treated with the same level of confidentiality as other medical records and that individuals are not penalized based on their genetic makeup. Additionally, public awareness campaigns and education initiatives are necessary to promote understanding and acceptance of genetic diversity.

As we continue to unlock the secrets of life through genomics, it is crucial that we navigate the social implications of this knowledge responsibly. By addressing genetic discrimination and its consequences, we can ensure that the genomic revolution benefits everyone, regardless of their genetic makeup. Ultimately, it is through ethical practices, legislation, and education that we can build a future where genetic information is used to empower individuals rather than discriminate against them.

Bioethics and Responsible Research

In the rapidly advancing field of genetics and genomics, the importance of bioethics and responsible research cannot be overstated. As we delve deeper into the secrets of life, it becomes imperative to establish ethical guidelines and ensure that scientific discoveries are utilized in a responsible manner that benefits everyone.

Bioethics is the study of ethical issues emerging from advances in biology and medicine. It provides a framework for evaluating the moral implications of scientific research and its applications. Responsible research, on the other hand, encompasses the ethical conduct of research, including considerations of transparency, integrity, and the impact on individuals and society.

One of the key areas where bioethics plays a crucial role in genetics and genomics is in the realm of privacy and informed consent. With the advent of technologies such as whole-genome sequencing, it is now possible to obtain an individual's complete genetic information. This raises concerns about the privacy and security of this sensitive data. Bioethics ensures that individuals have control over their genetic information and are adequately informed about the risks and benefits of genetic testing.

Another area of focus is the ethical use of gene editing technologies, such as CRISPR-Cas9. While these tools hold immense potential for curing genetic diseases, they also raise ethical concerns about altering the germline, which can have far-reaching consequences for future generations. Responsible research demands a careful evaluation of the risks and benefits, as well as broad societal consensus before employing such technologies.

Furthermore, bioethics guides the responsible conduct of research involving human subjects. It ensures that participants are treated with dignity and respect, and that their autonomy and welfare are protected. This includes obtaining informed consent, minimizing harm, and maintaining confidentiality.

In addition, responsible research acknowledges the importance of transparency and openness in scientific endeavors. It promotes the sharing of research findings, data, and methodologies, allowing for independent verification and replication. This fosters an environment of trust and accountability within the scientific community.

Ultimately, bioethics and responsible research are essential for harnessing the power of genetics and genomics in a way that benefits everyone. By upholding ethical principles, we can navigate the complex ethical dilemmas that arise and ensure that scientific advancements are used for the betterment of society. As we unlock the secrets of life, it is our collective responsibility to ensure that our actions are guided by compassion, integrity, and respect for the dignity of all individuals.

Chapter 6: The Future of Functional Genomics

Advancements in Technology and Data Analysis

In the realm of genetics and genomics, the advent of technology has revolutionized our understanding of life and has opened up a world of possibilities for everyone. The continuous advancements in technology and data analysis have propelled the genomic revolution, enabling scientists and researchers to unlock the secrets of life like never before.

One of the most significant advancements in technology has been the development of next-generation sequencing (NGS). This cutting-edge technique allows scientists to rapidly sequence the entire human genome, providing a wealth of information about an individual's genetic makeup. NGS has drastically reduced the time and cost required for genome sequencing, making it more accessible to researchers and healthcare professionals. As a result, personalized medicine has become a reality, as individuals can now receive tailored treatments based on their genetic profiles.

In addition to NGS, advancements in data analysis have also played a crucial role in the genomic revolution. As the volume of genomic data continues to grow exponentially, there is a pressing need for efficient and accurate methods of analyzing this vast amount of information. Bioinformatics, a field that combines biology and computer science, has emerged as a key discipline in making sense of genomic data. Through sophisticated algorithms and machine learning techniques, bioinformaticians can identify patterns, uncover disease-causing mutations, and predict an individual's risk of developing certain conditions.

Furthermore, the integration of artificial intelligence (AI) has further propelled the field of genomics. AI algorithms can process and analyze large datasets with unparalleled speed and accuracy, enabling researchers to make groundbreaking discoveries in a fraction of the time it would take using traditional methods. AI can also assist in drug discovery by simulating the effects of potential compounds on specific genetic targets, expediting the development of new treatments and therapies.

The advancements in technology and data analysis have not only impacted the field of genetics but have also had far-reaching implications for individuals. Genetic testing has become more accessible and affordable, allowing people to gain insights into their ancestry, identify potential health risks, and make informed decisions about their lifestyle and healthcare. Furthermore, the genomic revolution has the potential to revolutionize agriculture, conservation, and forensic science, among other fields.

As technology continues to evolve at an unprecedented pace, the future of genetics and genomics holds immense promise. The advancements in technology and data analysis have paved the way for a deeper understanding of life's complexities, offering new avenues for research, diagnosis, and treatment. Whether you are a scientist, healthcare professional, or simply interested in the wonders of life, the genomic revolution has something to offer everyone.

Integration of Functional Genomics with Artificial Intelligence

In recent years, there has been a rapidly growing interest in the field of genomics, where researchers are unlocking the secrets of life by studying the genetic makeup of organisms. This exciting field has paved the way for groundbreaking discoveries and advancements in various areas, including medicine, agriculture, and environmental science. However, with the sheer volume of genomic data being generated, there is a need for innovative tools and technologies to effectively analyze and interpret this wealth of information. This is where the integration of functional genomics with artificial intelligence (AI) comes into play.

Functional genomics aims to understand the functions and interactions of genes within a genome. It involves studying how genes are expressed, regulated, and interact with each other and the environment. On the other hand, AI refers to the development of computer systems that can perform tasks that typically require human intelligence, such as learning, reasoning, and problem-solving. By combining these two fields, researchers can leverage the power of AI algorithms to analyze vast amounts of genomic data and extract meaningful insights.

The integration of functional genomics with AI has the potential to revolutionize the field of genetics and genomics. It can accelerate the discovery of disease-causing genes, identify potential drug targets, and predict the effectiveness of different treatment options. For example, AI algorithms can analyze genomic data from patients with a particular disease and identify common genetic variations associated with the condition. This information can then be used to develop personalized treatment plans and improve patient outcomes.

Furthermore, AI can help uncover hidden patterns and connections within genomic data that may not be apparent to human researchers. This can lead to the discovery of novel gene functions, pathways, and regulatory mechanisms, shedding light on the complexity of life at a molecular level. By integrating functional genomics with AI, researchers can gain a deeper understanding of the genetic basis of diseases, as well as the fundamental biological processes that underlie life.

In conclusion, the integration of functional genomics with artificial intelligence holds tremendous potential for advancing our understanding of genetics and genomics. By harnessing the power of AI algorithms, researchers can analyze and interpret large-scale genomic data more efficiently and accurately. This integration can lead to groundbreaking discoveries, personalized medicine, and improved outcomes in various fields, benefiting not only scientists and researchers but also society as a whole. The genomic revolution is well underway, and the integration of functional genomics with AI is set to unlock even more secrets of life for everyone.

Potential Impacts on Healthcare and Society

In the era of the genomic revolution, the world is witnessing remarkable advancements and breakthroughs in the field of genetics and genomics. These discoveries have the potential to revolutionize healthcare and society as a whole, bringing about significant changes in the way we understand and approach various aspects of our lives.

One of the most significant potential impacts of genomics on healthcare is the ability to personalize medicine. With a deeper understanding of the human genome, healthcare professionals can develop tailored treatments and medications for individuals based on their genetic makeup. This personalized approach has the potential to enhance the effectiveness of treatments, minimize side effects, and ultimately lead to improved patient outcomes.

Moreover, genomics has the potential to transform the prevention and diagnosis of diseases. By analyzing an individual's genetic information, healthcare providers can identify individuals who are at a higher risk of developing certain diseases, allowing for targeted preventive measures. Additionally, genomics can enable early detection and more accurate diagnosis of diseases, enabling timely interventions and potentially saving lives.

The impact of genomics extends beyond healthcare, influencing various aspects of society. For instance, advancements in genetic testing and screening technologies have made it possible for individuals to gain insights into their ancestry and genetic heritage. This has not only fostered a sense of connection and belonging but also allowed people to better understand their predispositions to certain diseases, empowering them to make informed lifestyle choices and take proactive steps towards their well-being.

Furthermore, genomics has the potential to revolutionize the field of agriculture. By analyzing the genomes of crops and livestock, scientists can develop more resilient and disease-resistant varieties, ensuring food security and sustainability. Additionally, genomics can aid in the development of genetically modified organisms (GMOs) that are more nutritious and resistant to pests, potentially addressing global challenges such as malnutrition and food scarcity.

While the potential impacts of genomics on healthcare and society are vast and promising, it is crucial to address ethical, legal, and social implications (ELSI) to ensure responsible and equitable implementation. Ensuring privacy, preventing discrimination, and promoting equal access to genomic information and technologies are essential considerations to mitigate potential risks and ensure that the benefits are accessible to everyone.

In conclusion, the genomic revolution has the potential to transform healthcare and society as we know it. From personalized medicine to disease prevention, from understanding our genetic heritage to revolutionizing agriculture, genomics holds immense promise. However, it is crucial to navigate the associated ethical and societal challenges to ensure that these advancements benefit everyone and create a more equitable future.

Chapter 7: Accessible Functional Genomics for Everyone

Genomic Literacy and Education

In the ever-evolving world of genetics and genomics, it is becoming increasingly important for individuals from all walks of life to have a basic understanding of these fields. Genomic literacy is the key to unlocking the secrets of life and empowering everyone to make informed decisions about their health, their families, and their future.

Education plays a crucial role in ensuring that individuals are equipped with the knowledge and skills necessary to navigate the genomic revolution. Whether you are a healthcare professional, a student, a parent, or simply someone interested in delving into the fascinating world of genetics, acquiring genomic literacy is essential.

At its core, genomic literacy involves understanding the fundamentals of genetics and genomics, including DNA, genes, chromosomes, and the relationship between genotype and phenotype. It encompasses the ability to interpret genetic information, such as genetic test results, and to comprehend the implications of this information for individuals and their families.

One of the first steps towards genomic literacy is recognizing the importance of genetics in our lives. From inherited diseases to personalized medicine, genetics and genomics have a profound impact on our health and well-being. By understanding the basics of genetics, individuals can better comprehend the risks and benefits associated with genetic testing, explore their ancestry, and make informed decisions about reproductive choices.

To promote genomic literacy, educational initiatives must be accessible to everyone. This means breaking down complex scientific concepts into easily understandable terms and providing resources that cater to individuals with varying levels of scientific knowledge. Online courses, workshops, and educational materials can all contribute to enhancing genomic literacy among diverse audiences.

Furthermore, fostering genomic literacy should begin at an early age. Introducing genetics and genomics in school curricula can help young minds grasp the fundamental concepts and spark their interest in these fields. By nurturing a generation of scientifically literate individuals, we can ensure that the benefits of the genomic revolution are accessible to all.

In conclusion, genomic literacy and education are vital components of the genomic revolution. By equipping individuals with the knowledge and skills to understand and interpret genetic information, we empower them to make informed decisions about their health and embrace the opportunities that genomics present. From healthcare professionals to curious individuals, genomic literacy is a tool that can unlock the secrets of life for everyone. Let us embrace this knowledge and embark on a journey of discovery, understanding, and empowerment.

Open Access Data and Resources

In today's era of advanced technology and interconnectedness, the field of genetics and genomics has witnessed a remarkable shift towards open access data and resources. This subchapter aims to explore the significance of open access in unlocking the secrets of life for everyone, irrespective of their background or expertise. By breaking down barriers and fostering collaboration, open access has revolutionized the field, accelerating scientific progress and democratizing genomic knowledge.

Open access refers to the principle that scientific research, including genomic data and resources, should be freely accessible to the public. It enables researchers, clinicians, educators, and curious individuals from all walks of life to access, use, and build upon the vast wealth of genomic information available. This open approach encourages transparency, innovation, and the rapid dissemination of knowledge, ultimately benefiting society as a whole.

One of the key advantages of open access data is its potential to catalyze groundbreaking discoveries and advancements in genetics and genomics. By providing researchers with unrestricted access to a wealth of genomic data, open access initiatives have facilitated the identification of novel genes, genetic variations, and disease mechanisms. This not only enhances our understanding of the fundamental principles governing life but also paves the way for the development of personalized medicine, targeted therapies, and improved genetic counseling.

Open access resources have also played a pivotal role in fostering collaboration and interdisciplinary research. By freely sharing data, tools, and methodologies, researchers across the globe can collaborate and leverage each other's expertise to solve complex genetic puzzles. This collaborative environment promotes diversity of thought, accelerates the pace of discovery, and ensures that scientific advancements are not limited to a select few but are accessible to everyone, regardless of geographical location or financial constraints.

Moreover, open access data and resources empower individuals to take control of their own genomic information. In an era where direct-to-consumer genetic testing is becoming increasingly popular, open access allows individuals to access and interpret their own genetic data. This knowledge enables them to make informed decisions about their health, lifestyle choices, and even contribute to scientific research by voluntarily sharing their anonymized data.

In conclusion, open access data and resources have revolutionized the field of genetics and genomics by breaking down barriers and fostering collaboration. By democratizing access to genomic information, open access initiatives have accelerated scientific progress, enabled groundbreaking discoveries, and empowered individuals. As we continue to unravel the secrets of life, embracing open access is essential to ensure that the benefits of genomic research are accessible to everyone, regardless of their background or expertise.

Citizen Science and Participatory Genomics

In recent years, there has been a paradigm shift in the field of genetics and genomics. Where once these scientific domains were monopolized by specialized researchers in well-funded laboratories, today, ordinary individuals are actively participating in the pursuit of scientific knowledge. This chapter explores the fascinating world of citizen science and participatory genomics, highlighting the potential it holds for revolutionizing the way we understand and interact with our own genetic makeup.

Citizen science refers to the involvement of non-professional scientists, or "citizen scientists," in the scientific process. With the rise of technology and the accessibility of genetic testing, more and more people are engaging in activities that contribute to genetic research. This involvement can take various forms, such as participating in data collection, analyzing genomic data, or even generating new research questions. The democratization of genomics through citizen science has the power to accelerate scientific discoveries, provide diverse perspectives, and engage the public in meaningful ways.

Participatory genomics takes citizen science a step further by involving individuals in the analysis and interpretation of their own genetic information. With the advent of direct-to-consumer genetic testing, more people have access to their own genomic data than ever before. This presents a unique opportunity for individuals to contribute to scientific research by sharing their data and insights. By combining their genetic information with that of other participants, scientists can uncover patterns, identify new genetic variants, and gain a deeper understanding of the genetic basis of various traits and diseases.

The benefits of citizen science and participatory genomics extend beyond scientific discovery. They also empower individuals to take control of their own health and make informed decisions. By actively participating in the research process, individuals gain a better understanding of their own genetic makeup and how it relates to their health and well-being. This knowledge can drive personalized medicine, allowing for targeted treatments and interventions tailored to an individual's specific genetic profile.

However, while citizen science and participatory genomics offer immense potential, they also raise important ethical and privacy concerns. It is crucial to ensure that participants' data is protected and used responsibly, with informed consent and strict adherence to ethical guidelines.

In conclusion, citizen science and participatory genomics have the power to transform the field of genetics and genomics. By engaging the public in scientific research, we can unlock the secrets of life together, democratize access to genetic information, and empower individuals to take charge of their own health. It is an exciting time to be part of the genomic revolution, as we all have the opportunity to contribute to scientific progress and shape the future of genetic research.

Chapter 8: Challenges and Limitations in Functional Genomics

Data Interpretation and Analysis

In the fascinating realm of genetics and genomics, data interpretation and analysis play a pivotal role in unlocking the secrets of life. As we delve into the world of DNA, it becomes clear that the massive amount of information encoded within our genes requires careful analysis to extract valuable insights. This subchapter aims to demystify the process of data interpretation and analysis, making it accessible to everyone interested in the genomic revolution.

At its core, data interpretation involves making sense of the vast amounts of genetic data generated through advanced sequencing technologies. This data includes information about individual genes, their variations, and their interactions with other genes. By leveraging statistical techniques and computational tools, scientists can uncover patterns, correlations, and trends within the data, shedding light on the inner workings of our biology.

One key aspect of data interpretation is identifying genetic variations that correlate with specific traits or diseases. Through genome-wide association studies, scientists can sift through massive datasets to pinpoint genetic markers associated with conditions like cancer, heart disease, or even behavioral traits such as intelligence. This knowledge not only enhances our understanding of the genetic basis of these conditions but also paves the way for personalized medicine and targeted therapies.

Data analysis, on the other hand, involves manipulating and visualizing genetic data to extract meaningful information. Sophisticated algorithms and software programs aid in detecting patterns, identifying genes with similar functions, and predicting the impact of genetic variations. By harnessing the power of data analysis, researchers can develop new diagnostic tools, predict disease risks, and even design novel therapies tailored to an individual's unique genetic makeup.

Moreover, data interpretation and analysis are not limited to the realm of research laboratories. In the era of direct-to-consumer genetic testing, individuals can now access their personal genetic data. However, understanding the implications of this data requires proper interpretation. This subchapter will provide practical tips and guidelines on how to interpret and analyze personal genetic information responsibly, empowering readers to make informed decisions about their health and lifestyle choices.

In conclusion, data interpretation and analysis are crucial components of the genomic revolution, enabling us to unravel the mysteries of life encoded within our genes. By understanding the principles and techniques of data interpretation, individuals from all walks of life can actively participate in this scientific revolution, contributing to the advancement of genetics and genomics and ultimately improving human health and well-being for everyone.

Technical and Financial Constraints

In the ever-evolving field of genetics and genomics, there are numerous technical and financial constraints that researchers and scientists face. These obstacles not only hinder the progress of their work but also limit the accessibility of genomic information to the wider audience. Understanding these constraints is crucial to appreciating the challenges faced by scientists and the importance of ongoing research in the field.

One of the primary technical constraints in genomics is the complexity of the human genome itself. The human genome consists of billions of base pairs, and deciphering this intricate code requires sophisticated technology and expertise. Sequencing the entire genome is a time-consuming and expensive process, making it unattainable for many researchers and institutions. Furthermore, analyzing the vast amount of data generated from sequencing is a daunting task, requiring powerful computational resources and skilled bioinformaticians.

Financial constraints also pose significant challenges in genomics research. The cost of sequencing a genome has decreased significantly over the years, thanks to advancements in technology. However, it still remains a substantial investment, especially when multiple genomes need to be sequenced for comparative studies or large-scale population studies. Additionally, the ongoing maintenance and upgrading of expensive laboratory equipment, such as sequencers and high-performance computers, add to the financial burden.

These technical and financial constraints not only limit the pace of scientific discoveries but also impact the democratization of genomics.

The high costs associated with genomic research often lead to a lack of diversity in study populations, as researchers may focus on specific demographics that are more readily accessible or affordable. This can lead to biased findings and hinder the development of personalized medicine that benefits all individuals.

However, despite these challenges, there is hope on the horizon. Technological advancements continue to drive down the cost of sequencing, making it more accessible to researchers worldwide. Additionally, collaborative efforts between academia, industry, and governments are essential in securing funding and resources for genomics research. By sharing expertise, data, and resources, researchers can overcome technical and financial constraints, enabling them to unlock the secrets of life and make significant strides in the field of genetics and genomics.

In conclusion, technical and financial constraints are significant challenges faced by researchers in the field of genetics and genomics. These constraints limit the accessibility of genomic information and hinder scientific progress. However, with continued advancements in technology and collaborative efforts, researchers can overcome these obstacles, leading to a more inclusive and impactful genomic revolution that benefits everyone.

Ethical and Regulatory Challenges

In the realm of genetics and genomics, the rapid advancements in technology have ushered in a new era of possibilities and opportunities. However, with these exciting prospects come numerous ethical and regulatory challenges that must be carefully addressed. As we delve deeper into the genomic revolution, it is crucial to consider the implications and potential consequences of our actions.

One of the key ethical challenges surrounds the issue of privacy and data protection. As we gather vast amounts of genetic information, it is imperative that we establish robust safeguards to ensure the privacy and confidentiality of individuals. Genomic data is incredibly personal, containing sensitive information about an individual's health, ancestry, and predispositions to certain conditions. Therefore, it is essential to develop and enforce stringent regulations that protect the privacy rights of individuals while still allowing for scientific advancements.

Another ethical concern is the potential for genetic discrimination. Genetic information has the power to reveal predispositions to certain diseases or conditions, which could potentially be used by insurance companies or employers to discriminate against individuals. To prevent such abuses, it is essential to establish comprehensive legal frameworks that prohibit genetic discrimination and ensure equal access to healthcare and employment opportunities for all.

Additionally, the rapid pace of technological advancements raises questions about the potential misuse of genetic engineering and manipulation. While gene editing tools like CRISPR have the potential to revolutionize medicine, there is also the risk of using this technology for unethical purposes, such as creating "designer babies" or enhancing human traits beyond what is considered ethical. Therefore, it is crucial to establish clear guidelines and ethical boundaries to prevent the misuse of these powerful tools.

Regulatory challenges also arise in the context of intellectual property rights and access to genomic data. As discoveries in genetics and genomics continue to unfold, questions arise about who owns the rights to genetic information and how it can be used. Balancing the need for commercial incentives with the goal of advancing scientific knowledge is a complex task that requires careful consideration and regulation.

In conclusion, the genomic revolution has brought forth tremendous possibilities for unlocking the secrets of life. However, it is essential to address the ethical and regulatory challenges that accompany these advancements. By establishing robust privacy protections, preventing genetic discrimination, setting ethical boundaries for genetic engineering, and addressing issues of intellectual property and data access, we can ensure that the genomic revolution benefits everyone while upholding the highest ethical standards. It is our responsibility as a society to navigate these challenges with care, balancing scientific progress and human values to create a future where genetics and genomics are used for the betterment of all.

Chapter 9: Case Studies in Functional Genomics

Unraveling Genetic Causes of Complex Diseases

In the realm of genetics and genomics, one particular area of study has captivated researchers and scientists worldwide - the unraveling of genetic causes behind complex diseases. This subchapter aims to shed light on the groundbreaking research and discoveries made in this field, providing insight into the intricate web of genes and their role in the development of multifaceted diseases.

Complex diseases, such as diabetes, cancer, Alzheimer's, and heart disease, have long puzzled scientists due to their multifactorial nature. Unlike single-gene disorders that are caused by mutations in a single gene, complex diseases are influenced by a combination of genetic, environmental, and lifestyle factors. Understanding the genetic component of these diseases has been a formidable challenge, but recent advancements in genomics have brought us closer to unraveling the complex genetic architecture behind them.

Genome-wide association studies (GWAS) have been instrumental in identifying common genetic variants associated with complex diseases. By analyzing the genomes of thousands of individuals, researchers can pinpoint specific genetic variations that increase the risk of developing a particular disease. These findings have not only provided valuable insights into the underlying genetic causes but have also paved the way for personalized medicine, allowing healthcare professionals to tailor treatments based on an individual's genetic profile.

Furthermore, advancements in sequencing technologies have enabled researchers to delve deeper into the human genome, uncovering rare genetic variants that may contribute to complex diseases. Whole-exome sequencing, for instance, focuses on decoding the protein-coding regions of the genome, where disease-causing mutations are more likely to be found. This approach has proven successful in identifying rare genetic variants responsible for certain types of cancer and neurological disorders.

Moreover, the integration of genomics with other -omics disciplines, such as transcriptomics and proteomics, has provided a comprehensive picture of the molecular mechanisms underlying complex diseases. By studying how genes are expressed and how their proteins function within cells, researchers can gain a deeper understanding of disease pathways and potential therapeutic targets.

While the unraveling of genetic causes of complex diseases has undoubtedly expanded our knowledge, challenges still remain. The interaction between genes and the environment, along with the ethical implications of genetic testing and privacy concerns, necessitate ongoing research and careful consideration.

In conclusion, the quest to unravel the genetic causes of complex diseases has revolutionized the field of genetics and genomics. Through GWAS, sequencing technologies, and integrative approaches, scientists have made significant strides in identifying the genetic variants that contribute to these multifaceted conditions. As our understanding of the complex interplay between genes and diseases continues to evolve, so too does the potential for personalized treatments and improved healthcare outcomes for everyone.

Leveraging Functional Genomics for Precision Agriculture

In recent years, the field of functional genomics has emerged as a game-changer in the realm of agriculture. With the world's population steadily increasing, the demand for food production has skyrocketed, posing significant challenges for farmers and agricultural scientists. However, the integration of functional genomics into precision agriculture offers a promising solution to these challenges, revolutionizing the way we cultivate crops and ensuring sustainable food production for the future.

Functional genomics involves the study of how genes function and interact within an organism. By understanding the genetic makeup of plants, scientists can uncover valuable insights into their growth, development, and response to environmental factors. This knowledge can be harnessed to optimize agricultural practices, enhance crop yields, and mitigate the impact of pests, diseases, and climate change.

Precision agriculture, on the other hand, refers to the practice of using advanced technologies and data analytics to make informed decisions regarding crop management. By combining functional genomics with precision agriculture, farmers can customize their approach to suit the specific needs of each crop, maximizing productivity while minimizing resource wastage.

One of the key applications of functional genomics in precision agriculture is the identification and breeding of genetically superior crop varieties. By studying the genes responsible for desirable traits such as drought tolerance, disease resistance, and higher yields, scientists can develop new crop varieties that are better equipped to thrive in challenging environments. This not only enhances food security but also reduces the need for chemical inputs, resulting in more sustainable and environmentally friendly farming practices.

Another area where functional genomics has proved invaluable is in the development of precision pest management strategies. By understanding the genetic mechanisms of pest resistance and susceptibility, scientists can devise targeted approaches to control pests, reducing the reliance on broad-spectrum pesticides. This not only minimizes the impact on beneficial insects and the environment but also helps farmers save costs by using only the necessary amount of chemicals.

Furthermore, functional genomics has enabled the development of precision nutrient management systems. By analyzing the genetic makeup of plants, scientists can determine their nutrient requirements and tailor fertilizer application accordingly. This ensures that crops receive the optimal amount of nutrients, reducing fertilizer runoff and minimizing environmental pollution.

In conclusion, the integration of functional genomics into precision agriculture holds immense potential for revolutionizing the way we produce food. By harnessing the power of genetics and genomics, farmers can cultivate genetically superior crops, implement targeted pest management strategies, and optimize nutrient usage. This not only enhances productivity and profitability but also promotes sustainable and environmentally friendly agricultural practices. As we embark on the genomic revolution, functional genomics will play a pivotal role in unlocking the secrets of life for everyone involved in the field of genetics and genomics, ultimately benefiting the entire population.

Conservation Genomics: Protecting Endangered Species

In the face of accelerated habitat destruction, climate change, and human activities, the world is experiencing an alarming decline in biodiversity. Numerous species are on the brink of extinction, with their genetic diversity at risk of being lost forever. However, amidst this crisis, a powerful tool has emerged - conservation genomics - offering hope for the preservation and protection of endangered species.

Conservation genomics is a rapidly advancing field that combines the power of genetics and genomics to understand the genetic makeup of endangered species and develop effective conservation strategies. By unraveling the secrets hidden within the DNA of these species, scientists can gain valuable insights into their population dynamics, evolutionary history, and adaptability to changing environments.

One of the key applications of conservation genomics is identifying and preserving genetic diversity. Genetic diversity is crucial for a species' ability to adapt and survive in the face of changing conditions. Through advanced genomic techniques, scientists can analyze the DNA of endangered species and identify areas of low genetic diversity. This information helps conservationists prioritize specific populations for protection and identify individuals for captive breeding programs, ensuring the preservation of genetic diversity for the future.

Conservation genomics also enables researchers to study the impacts of habitat fragmentation and climate change on endangered species. By comparing the genomes of different populations, scientists can measure the extent of genetic differentiation resulting from habitat loss. This knowledge allows for the development of targeted conservation strategies, such as the creation of wildlife corridors to reconnect fragmented habitats and facilitate gene flow between isolated populations.

Furthermore, conservation genomics plays a vital role in combating illegal wildlife trade. By using genomic tools, scientists can identify the geographic origin of confiscated specimens and trace their routes through complex smuggling networks. This information not only aids law enforcement agencies in prosecuting wildlife traffickers but also helps conservationists target areas where increased protection measures are necessary.

The importance of conservation genomics extends beyond individual species to entire ecosystems. By understanding the genetic diversity and interactions of different species within an ecosystem, scientists can assess its resilience and predict how it may respond to future challenges. This knowledge allows for the development of proactive conservation measures that safeguard the delicate balance of ecosystems and promote their long-term sustainability.

In conclusion, conservation genomics represents a powerful tool in the fight against species extinction. By harnessing the knowledge provided by genetics and genomics, scientists can devise targeted and effective strategies to protect endangered species and preserve the genetic diversity essential for their survival. The application of genomic techniques not only aids in understanding the challenges faced by endangered species but also provides hope for a future where biodiversity thrives and our planet's delicate ecosystems are preserved for generations to come.

Chapter 10: Conclusion

Recap of Key Concepts

In this subchapter, we will provide a brief recap of the key concepts covered in our book, "The Genomic Revolution: Unlocking the Secrets of Life for Everyone." Whether you are new to the field of genetics and genomics or have been following the advancements in this area, this summary will serve as a helpful reminder and reference point for the information presented.

Throughout the book, we have explored the incredible potential of genomics to transform our understanding of life and revolutionize various fields, from healthcare to agriculture. We began by introducing the basics of genetics, explaining how DNA serves as the blueprint for life and how it is organized into genes and chromosomes.

We then delved into the technologies that have enabled the genomic revolution, such as DNA sequencing and gene editing tools like CRISPR-Cas9. These breakthroughs have allowed scientists to decipher the entire human genome and uncover the genetic variations that contribute to diseases and traits.

Next, we explored the implications of genomics in healthcare. By understanding the genetic basis of diseases, healthcare providers can now offer personalized medicine, tailoring treatments to individual patients based on their genetic makeup. This approach has the potential to revolutionize how we prevent, diagnose, and treat diseases, leading to improved outcomes and reduced healthcare costs.

Furthermore, we discussed the ethical considerations surrounding genomics, including privacy concerns and the potential for discrimination based on genetic information. As genomics becomes more accessible to the general public, it is crucial to address these ethical issues and establish guidelines to protect individuals' rights and ensure equitable access to genomic information.

In the field of agriculture, genomics offers tremendous opportunities to enhance crop yields, develop disease-resistant plants, and improve livestock breeding. By understanding the genetic makeup of different organisms, scientists can optimize agricultural practices and create sustainable solutions to feed a growing population.

Finally, we emphasized the importance of genomics literacy for everyone. Understanding the fundamental concepts of genetics and genomics empowers individuals to make informed decisions about their health, engage in discussions about genetic research, and advocate for policies that promote the responsible use of genomic information.

In conclusion, "The Genomic Revolution: Unlocking the Secrets of Life for Everyone" has provided a comprehensive overview of the key concepts in genetics and genomics. By embracing this knowledge, individuals from all walks of life can actively participate in and benefit from the genomic revolution, ultimately shaping a future where genomics is accessible and beneficial for everyone.

The Impact of Functional Genomics on Society

Functional genomics has emerged as a transformative field of study within the broader realm of genetics and genomics, revolutionizing our understanding of life and its intricate mechanisms. This subchapter explores the profound impact that functional genomics has had on society as a whole, shedding light on its implications for individuals, healthcare, and scientific advancements.

At its core, functional genomics aims to decipher the functionality and interactions of genes within an organism. By investigating how genes are expressed and regulated, scientists have unlocked a wealth of knowledge that has far-reaching consequences for society. One major outcome of functional genomics is the ability to identify disease-causing genes and develop targeted therapies, paving the way for personalized medicine. This breakthrough has transformed the landscape of healthcare, allowing for more accurate diagnoses, tailored treatment plans, and improved patient outcomes.

Additionally, functional genomics has revolutionized our understanding of complex diseases, such as cancer and neurological disorders, by revealing the underlying molecular mechanisms. This knowledge has not only facilitated the development of novel therapies but also enabled the identification of biomarkers for early detection and prevention. Consequently, functional genomics has the potential to save countless lives and alleviate the burden on healthcare systems worldwide.

Beyond healthcare, functional genomics has also had significant implications for agriculture and food production. By studying the genes of crops and livestock, scientists can identify desirable traits, such as disease resistance and increased yield, leading to the development of genetically modified organisms (GMOs). While GMOs remain a topic of debate, functional genomics has the potential to address food security challenges and mitigate the effects of climate change on agriculture.

Moreover, functional genomics has fueled scientific advancements and breakthroughs in other disciplines. It has facilitated the study of evolutionary biology, allowing researchers to explore the genetic basis of species diversification and adaptation. Functional genomics has also enabled the development of synthetic biology, where scientists manipulate and engineer genes to create new organisms or modify existing ones for various purposes, including biofuel production and environmental remediation.

As functional genomics continues to advance, it is crucial to address its ethical, legal, and social implications. Issues such as privacy concerns, genetic discrimination, and equitable access to genomic technologies must be carefully considered to ensure that the benefits of functional genomics are accessible to all and do not exacerbate existing societal disparities.

In conclusion, functional genomics has had a profound impact on society, transforming healthcare, agriculture, and scientific advancements. Its ability to unravel the secrets of life at the molecular level has revolutionized our understanding of genetics and genomics, opening up new possibilities for personalized medicine, disease prevention, and sustainable food production. However, as we navigate the genomic revolution, it is essential to address the ethical and social implications to ensure that the benefits are shared by everyone and contribute to a more equitable and inclusive society.

Looking Ahead: What Lies Beyond the Genomic Revolution?

As we stand on the precipice of the genomic revolution, it is natural to wonder what lies beyond this groundbreaking era of genetics and genomics. The progress we have witnessed in decoding the secrets of life has been nothing short of remarkable, but it is only the beginning. The future holds even more exciting possibilities, promising to reshape our understanding of the world and our place in it.

One area that shows tremendous promise is personalized medicine. With the advent of genomics, we can now better understand the genetic variations that contribute to disease susceptibility and treatment response. This knowledge opens the door to a new era of precision medicine, where treatments can be tailored to an individual's unique genetic makeup. This not only enhances the efficacy of treatment but also minimizes side effects and improves patient outcomes. In the coming years, we can expect personalized medicine to become more accessible and widespread, revolutionizing healthcare as we know it.

Another area where genomics is set to make a significant impact is in agriculture. As the global population continues to grow, the demand for food is increasing exponentially. Genomic technologies can help address this challenge by improving crop yields, enhancing resistance to diseases, and increasing nutritional content. By using genomics to develop genetically modified crops, we can produce more food with fewer resources, reducing the environmental impact of agriculture and ensuring food security for future generations.

Beyond the practical applications, the genomic revolution will continue to unravel the mysteries of life. As we delve deeper into the complexities of our genetic code, new discoveries will emerge, shedding light on the fundamental principles that govern life itself. We will gain a deeper understanding of how genes interact with each other and the environment, unraveling the intricate web of life's processes. This knowledge will not only fuel further scientific advancements but also challenge our perceptions of what it means to be human.

However, we must also acknowledge that the genomic revolution raises ethical questions that must be addressed. As we gain greater control over our genetic destiny, we must grapple with issues of privacy, consent, and equitable access to genomic information. The responsible use of genomics demands careful consideration of these ethical considerations, ensuring that the benefits of the revolution are enjoyed by all.

In conclusion, the genomic revolution has laid a foundation for transformative advancements in genetics and genomics. Looking ahead, we can anticipate a future where personalized medicine, sustainable agriculture, and a deeper understanding of life's mysteries become a reality. However, as we forge ahead, we must do so with a keen awareness of the ethical implications and strive for a future where the benefits of the genomic revolution are accessible to everyone.